Guide

to

Pre - 1930

AIRCRAFT ENGINES

compiled by

M. S. Rice

PUBLISHED in the U.S.A.
by

ISBN 0 – 87994 – 016 – 6

LIBRARY OF CONGRESS CATOLOG CARD NUMBER: 72 - 83187

FOREWORD

The throaty roar, the staccato drumming, the grinding growl, the clatter and popping–all are sounds reminiscent of bygone days in aviation, when the engine was every bit as exciting as the airplane it powered. These sounds of yesteryear, except in a few rare instances where the aircraft and engine are preserved and operated, are merely fond memories. We can't reproduce the sounds but we have reproduced the historical information about these engines and offer it to you in this GUIDE TO PRE-1930 AIRCRAFT ENGINES.

Perhaps no finer example of engineering can be found than is exhibited by the aircraft engine. It contains nothing that it does not require, and every single part of it must have a long service life and return the maximum possible output of power for its weight.

Reliability has always been of greatest importance and the aircraft engine of today can run almost indefinitely. Even in the "golden age" of aviation it had remarkable reliability for transoceanic flights and endurance attempts. The power-to-weight ratio was steadily improved as was the proportionate rate of fuel consumption.

The radial engine achieved a tremendously favorable power-to-weight ratio, but at the expense of frontal area. Consequently, cowling had to be provided to streamline and force the cooling, but it eliminated the weight of liquid-cooling systems along with their inherent problems.

In their respective developments, the airplane engine and automobile engine each took a different course. The weight of an aircraft engine had to be kept as light as possible, and the engine had to provide a wide range of performance under a wide range of operating conditions. On the other hand, the auto engine was never critical as to weight, and this fact in turn insured a degree of safety. The auto engine operates under far worse conditions than does the aircraft engine, and suffers much abuse at the hands of careless owners and operators, and it still continues to perform with matchless reliability. This is not so with the aircraft engine.

Progress in aircraft engines moved along the following lines:

1. Increased output for the unit of weight
2. Improved fuel economy
3. Minimization of frontal area
4. Improved reliability and engine life

This progress was realized through improvements in materials, design, fuels, and lubricants, in addition to accessory development.

The period where this GUIDE leaves off just saw the beginnings of practical application of supercharging and the variable-pitch propeller, either or both of which greatly improved the efficiency and output of the engine, but the engine itself was still pretty much the same. The variable-pitch propeller provided a wide spectrum of engine speeds without having to adjust the throttle, while supercharging effected a considerable range of power output and was controlled with the air/fuel mixture by the pilot in order to obtain the most efficient combination.

The internal-combustion engine, prior to the jet age, was the only prime mover that proved capable of meeting all the exacting requirements of aircraft power. The designer of an aircraft engine always had to consider these important basic requirements:

1. Reliability
2. Durability
3. Compactness
4. Low weight per horsepower
5. High specific output
6. Reasonable cost
7. High thermal efficiency
8. Freedom from vibration
9. Ease of maintenance in the field
10. Operating flexibility

As mentioned above, reliability was always considered to top the list because without it you have nothing. Each of the others contributed to reliability in one way or another. In the air, no item of the powerplant design or fabrication is unimportant.

Even greater changes have been made since 1930 in metallurgy and design—strengthening of parts and their reliability, and reduction in operating costs and maintenance. Military requirements placed great emphasis on ruggedness and reliability, but were little concerned with fuel-consumption ratio's. However, the power output of these engines toward the start of World War II increased several fold, so the increase in fuel consumption was more than offset by the power that was made available for military purposes.

The installation of engines, propellers, and accessories in earlier aircraft was very simple, and each related to the other in one way or another. Later engines, particularly the more powerful types, were quite complex and depended on such variables as power, propeller speeds, supercharging, fuel ratings, etc. The early pilot was mainly concerned with keeping the engine running long enough to get to where he was going. This gave way in the thirties and forties to the modern pilot extracting from his engine the maximum operating efficiency and performance within economy and safety limitations.

Where many of the earlier engines had an operating life of less than 75 hours, 1000 hours or better between major overhauls is not uncommon since World War II. An inverse relationship exists, however, between the power produced by an engine and the number of hours it can be expected to last, so higher power demands greatly reduce the time between engine overhauls.

Geometric forms of engines to a great extent have been dictated by how compact the unit must be, and this in turn was limited by the need to eliminate the heat through the cooling fins at a fast rate. Liquid-cooled engines sent the heat into the coolant and the latter through the radiator. While the system was heavier and more complicated, it did provide a better opportunity to streamline the airplane. All of the heat other than that turned into work had to be released ultimately into the surrounding air. To get rid of almost 75 percent of this heat through the cooling and exhaust systems and still have it compact was a major engineering feat.

By and large, the elimination of the liquid-cooling system improved the dependability of the aircraft engine, particularly the single-row radials, and the double-row radials that appeared in the middle thirties. Their dependability on long-distance flights is legend. The only application of liquid-cooled engines in latter years was in various fighter-interceptor aircraft which did not require expectation of long service life.

Radial reciprocating engines achieved approximately one pound per horsepower, but still could not compare with the efficiency of the jet. The aircraft engine, though still the standard in private aircraft manufactured today, has to be one of the marvels of engineering. Between the rotary-engine era of World War I and the twin-row radial and liquid-cooled V-12 of less than 20 years later, and the lightplane engines that followed, America opened the far corners of Earth to the airplane in peace and in war, and thereby changed the destiny of man.

The GUIDE TO PRE-1930 AIRCRAFT ENGINES is offered as testimony and a source of reference for the aircraft engines of that "golden age" of aviation. It brings together under one cover the details and important information concerning the great variety of American aircraft powerplants which contributed to America's tremendous achievements in aviation.

Because of the variety of combinations possible between engines and carburetors and propellers, each engine is covered only in its basic-design model. Consequently, performances, and to some extent weights, will vary depending on the combinations employed to realize a particular power requirement.

Michael S. Rice
Publisher

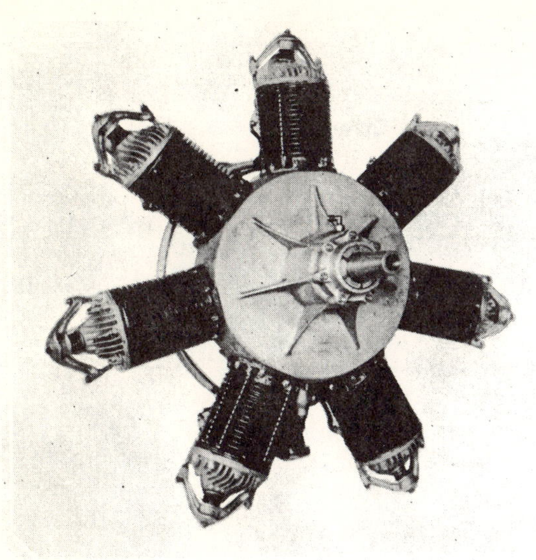

Name of Engine:	**SPEER S-2-C.**
Manufactured by:	Aero Development Co., Cedar Falls, Iowa.
Type:	7 cylinder, radial, air-cooled.
Rating:	120 H.P. at 1800 R.P.M.
Displacement:	491 cu. in.
Compression Ratio:	5.2 to 1.
Dimensions:	Length, overall ... 27⅜" Diameter overall ... 38½" Bore ... 4⅛" Stroke ... 5¼"
Weight:	330 lbs. (without hub or starter).
Fuel Consumption at Rated H.P.:	Not more than .53 lbs. per H.P. hr.
Oil Consumption:	Not more than .025 lbs. per H.P. hr.
Lubrication:	Pressure and scavenging gear pumps.
Ignition:	Dual Scintilla MN-7D
Carburetion:	Stromberg NAS_5B.
Spark Plugs:	2 per cyl. (Champion).
Price:	On application.

Accessories: (Cost Extra)

Eclipse Starter.
Viking Fuel Pump.
Tools.
Instruction Book.

Special Features:

Solid Master Rod Operating on Roller Bearings.
Two-piece crankshaft.

Cast Aluminum Cylinder head bolted to Nickel Iron barrel, Cylinder is of superposed or "F" head construction, the exhaust valve being in the head and the intake valve in the ell. The valves are operated by a special action that eliminates all parts outside the crankcase that require lubrication. The combustion chamber is of a type that provides high turbulence and resulting fuel economies.

All accessories are grouped at the rear of the engine for protection from the weather and to provide ease of maintenance.

Due to the fact that there are no babbit bearings, the need for high pressure lubrication has been eliminated, consequently only sufficient pressure is carried to serve as an indication on the panel that oil is flowing.

Name of Engine:	**A.C.E.**
Manufactured by:	Air Craft Engine Corp., Philadelphia, Pa.
Type:	7 cylinder, radial air-cooled
Rating:	140 H.P. at 1800 R.P.M.
Displacement:	528.8 cu. in.
Compression Ratio:	4.7 to 1.
Dimensions:	Length, overall ....33" Diameter, overall ....44" Bore ....4½" Stroke ....4¾"
Weight:	375 lbs.
Fuel Consumption at Rated H.P.:	Not more than .55 lbs. per H.P. hr.
Oil Consumption:	Not more than .017 lbs. per H.P. hr.
Lubrication:	1 pressure pump; 2 scavenger pumps.
Ignition:	2 Scintilla.
Carburetion:	NAR-5 Stromberg.
Spark Plugs:	2 per cyl.
Price:	On application.

Accessories: (Cost Extra)

Starter. Propeller Hub. Generator.

Special Features:

Permanent mould cast aluminum alloy cylinder head, bolted to forged steel barrel with 12 nickel steel bolts and nuts.

Open rocker arm system giving better valve and valve spring cooling.

High Tungsten—Vanadium inlet valves—C.N.S. special exhaust valves.

All shafts ball bearing.

All aluminum casting—heat treated alloys.

Pistons aluminum alloy heat treated, special ribbing; fuel floating pins fitted with aluminum alloy plugs.

Carburetor air heater with control.

Name of Engine:	**MURRAY-AJAX.**
Manufactured by:	Aircraft Holding Corp., Culver City, Calif.
Type:	6 cylinder, fixed radial, air cooled, two cycle, valveless, supercharged.
Commercial Rating:	80 H.P. at 1400 R.P.M.
Displacement:	484 cu. in.
Compression Ratio:	6 to 1.
Dimensions:	Length, overall ... 26" Diameter, overall ... 36" Bore ... 4⅜" Stroke ... 5⅜"
Weight:	210 lbs.
Fuel Consumption at Rated H.P.:	Not more than .065 gal. per H.P. hr.
Oil Consumption:	Not more than .04 lbs. per H.P. hr.
Lubrication:	Force feed, 3 outlet pump; scavenger pump.
Ignition:	Dual Scintilla.
Carburetion:	Zenith.
Spark Plugs:	2 per cyl. Bosch.
Price:	$1,800.

Accessories: (Cost Extra)

Propeller Hub. Starter.

Special Features:

No valves, cams or push rods—⅓ less parts. Supercharger with variable pressure of 1.5 to 10 lbs. per sq. in.

Cylinders are clamped between two halves of crankcase eliminating hold-down bolts.

Piston and cylinder heads can be scraped of carbon through exhaust ports without dismantling engine.

Smaller overall diameter and less wind resistance than a 4-stroke motor of equivalent power.

Decreased vibration due to fact that every cylinder delivers a power stroke to the shaft once every revolution instead of once every two revolutions as in the 4-stroke type.

Improved cooling due to release of flame at bottom of power stroke through large exhaust ports and added cooling effect of incoming charge scouring two opposite sides of cylinder wall.

Engine develops rated power at lower R.P.M. obviating necessity of gearing down the propeller and obtaining good engine life through slower operation.

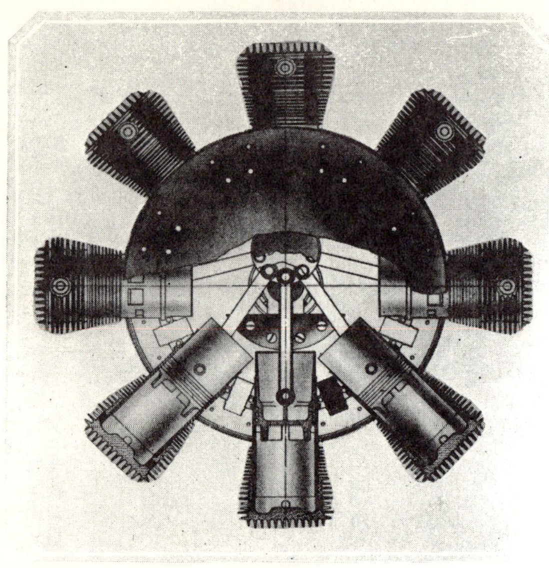

Name of Engine:	**MURRAY-ATLAS.**
Manufactured by:	Aircraft Holding Corp. Culver City, Calif.
Type:	8 cylinder, radial, air-cooled, 2 cycle, valveless, supercharged.
Commercial Rating:	120 H.P. at 1400 R.P.M.
Displacement:	646 cu. in.
Compression Ratio:	6 to 1.
Dimensions:	Length, overall 26" Diameter, overall 36" Bore 4⅜" Stroke 5⅜"
Weight:	260 lbs.
Fuel Consumption at Rated H.P.:	Not more than .065 gal. per H.P. hr.
Oil Consumption:	Not more than .04 lbs. per H.P. hr.
Lubrication:	Force feed, 3 outlet pump; scavenger pump.
Ignition:	Dual Scintilla.
Carburetion:	Zenith.
Spark Plugs:	2 per. cyl. Bosch.
Price:	$2,700.

Accessories: (Cost Extra)

Propeller Hub.

Starter.

Special Features:

No valves, cams or push rods, hence ⅓ less parts than ordinary engine.

Supercharger permits variation in fuel pressure of 1.5 to 10 lbs. per sq. in.

Cylinders are clamped between two halves of crankcase eliminating hold-down bolts.

Piston and cylinder heads can be scraped of carbon through exhaust ports without dismantling engine.

Smaller overall diameter and less wind resistance than a 4-stroke motor of equivalent power.

Decreased vibration due to fact that every cylinder delivers a power stroke to the shaft once every revolution instead of once every two revolutions as in the 4-stroke type.

Improved cooling due to release of flame at bottom of power stroke through large exhaust ports and added cooling effect of incoming charge scouring two opposite sides of cylinder wall.

Engine develops rated power at lower R.P.M. obviating necessity of gearing down the propeller and obtaining good engine life through slower operation.

Name of Engine:	**HESS "WARRIOR"**
Manufactured by:	Alliance Aircraft Corp., Alliance, Ohio.
Type:	7 cylinder, fixed radial, air cooled.
Military Rating:	115 H.P. at 1925 R.P.M.
Commercial Rating:	115 H.P. at 1925 R.P.M.
Displacement:	447 cu. in.
Compression Ratio:	5.2 to 1.
Dimensions:	Diameter, overall .... 37" Bore .... $4\frac{1}{4}$" Stroke .... $4\frac{1}{2}$"
Weight:	295 lbs.
Fuel Consumption at Rated H.P.:	Not more than .58 lbs. per H.P. hr.
Oil Consumption:	Not more than .025 lbs. per H.P. hr.
Lubrication:	Pressure and scavenger pumps.
Ignition:	2 Scintilla magnetos.
Carburetion:	Holley.
Spark Plugs:	2 per cyl. B. G.
Price:	$2,250.

Special Features:

Cylinder Construction—The aluminum alloy head is screwed and shrunk onto the steel barrel, making a unit which is gas tight, will not loosen, and affords excellent heat transference. Bronze inserts and bushings, also shrunk in, take care of all points of wear such as valve seats, valve guides and spark plug bushings.

Crankshaft and Rods—A two-piece crankshaft with the crank pin integral with the front half is used to transmit the power to the propeller. This construction permits the use of a one-piece master rod and the attendant higher speeds. The articulating rods are of H section and machined all over.

Bearings—The shaft is supported by two roller bearings and has a deep groove ball thrust bearing. Steel backed babbitt bearings are used for the master rod and cam.

The valve gear is entirely enclosed, keeping the lubricant in, and the dirt and grit out. This means a clearer and more economical engine and also less wear with the attendant replacement of parts.

All accessories are grouped at the rear of the engine, out of the slip stream, giving a clearer installation, case of control hook up and protection from the elements.

Name of Engine:	**AMERICAN CIRRUS.**
Manufactured by:	American Cirrus Engines, Inc., Marysville, Mich.
Type:	4 cylinder in line, vertical air cooled.
Rating:	90 H.P. at 1900 R.P.M. } 5.1 compression ratio 95 H.P. at 2100 R.P.M. 93 H.P. at 1900 R.P.M. } 5.4 compression ratio 98 H.P. at 2100 R.P.M.
Displacement:	310 cu. in.
Compression Ratio:	5.4 to 1 (max.)—5.1 to 1 (min.)
Dimensions:	Length, overall .... 38½" Height, overall .... 36⅜" Width, overall .... 18¾" Bore .... 4⅜" Stroke .... 5⅛"
Weight:	275 lbs. without hub or starter.
Fuel Consumption at Rated H.P.:	Not more than .54 lbs. per H.P. hr.
Oil Consumption:	Not more than .0213 lbs. per H.P. hr.
Lubrication:	Force feed, wet sump.
Ignition:	2 Scintilla magnetos.
Carburetion:	Stromberg.
Spark Plugs:	2 per cyl. B. G. and AC.
Price:	$1,600.

Accessories: (Cost Extra)
Starter.

Special Features:
Light Weight.

Name of Engine:	**HUDSON-HAWK.**
Manufactured by:	American Eagle Aircraft Corp., Kansas City, Mo.
Type:	6 cylinder, fixed radial, air-cooled, 4 cycle.
Rating:	100 H.P. at 1800 R.P.M.
Displacement:	372 cu. in.
Compression Ratio:	5.4 to 1.
Dimensions:	Length, overall .28" Diameter, overall .33" Bore 3.875" Stroke 5.250"
Weight:	326 lbs. without supercharger.
Fuel Consumption at Rated H.P.:	Not more than .50 lbs. per H.P. hr.
Oil Consumption:	Not more than .035 lbs. per H.P. hr.
Lubrication:	Force feed—pressure and scavenging pumps.
Ignition:	2 Scintilla.
Carburetion:	McCadden.
Spark Plugs:	2 per cyl.
Price:	On application.

Accessories: (Cost Extra)

Self-starter.

Supercharger.

Special Features:

Small diameter, L-head.
Solid (no cap) connecting rods.
Built up crankshaft.

All moving parts under force feed lubrication and fully enclosed.

Name of Engine:	**AXELSON B.**
Manufactured by:	Axelson Aircraft Engine Co., Box 337, Los Angeles, Calif.
Type:	7 cylinder, fixed radial, air cooled. Approved Dept. Commerce Certificate No. 16.
Rating:	150 H.P. at 1800 R.P.M.
Displacement:	612.3 cu. in.
Compression Ratio:	5 to 1.
Dimensions:	Length, overall ... 37¾″ Diameter, overall ... 45″ Bore ... 4½″ Stroke ... 5½″
Weight:	420 lbs. (without hub or starter).
Fuel Consumption at Rated H.P.:	Not more than .55 lbs. per H.P. hr.
Oil Consumption:	Not more than .017 lbs. per H.P. hr.
Lubrication:	Duplex gear pressure pump, 60-70 lbs., ½ crankshaft speed.
Ignition:	2 Scintilla Magnetos MN-7 D type.
Carburetion:	Stromberg NAR 5-A.
Spark Plugs:	1 per cyl.
Price:	$2,950.

Equipment:

Tool Kit.
Propeller Hub Nuts.
Exhaust Stacks.
Instruction Book.

Accessories: (Cost Extra)

Any type Starter.
Fuel Pump.
Ignition Switch.
Temperature Gauges.
Propeller.

Special Features:

Cylinder barrels of Ni-Cr Steel forgings, with cylinder head of cast aluminum alloy screwed and shrunk on to barrels. Cooling pins machined integral with barrel.

Hinged rocker boxes compensate for expansion due to heat, thereby automatically maintaining valve tappet clearance regardless of engine temperature.

Lubricating system used makes oil leaks impossible.

Pistons designed for maximum lubricating of upper cylinder walls, and minimum vibration.

Name of Engine:	**BLISS JUPITER.**
Manufactured by:	E. W. Bliss Company, Brooklyn, N. Y.
Type:	9 cylinder, radial, air cooled (Approved Type Certificate).
Rating:	525 H.P. at 1700 R.P.M.
Displacement:	1750 cu. in.
Compression Ratio:	5 to 1.
Dimensions:	Diameter, overall .... 54¾" Bore .... 5.75" Stroke .... 7.50"
Weight:	720 lbs.
Fuel Consumption at Rated H.P.:	Not more than .50 lbs. per H.P. hr.
Oil Consumption:	Not more than .035 lbs. per H.P. hr.
Lubrication:	Pressure pump, 40 lbs. per sq. in.
Ignition:	Dual G. E. Magnetos.
Carburetion:	Bristol Triplex.
Spark Plugs:	1 per cyl. K.L.G.
Price:	On application.

Accessories: (Cost Extra)
Eclipse Starter.

Special Features:

Supercharger.

2 to 1 Epicyclic reduction gear optional.

Cylinder barrels of steel, head of aluminum alloy, secured to barrels by studs and set screws.

Invar sleeves fitted under nuts to compensate for unequal expansion.

Pistons of cast Y-alloy (Nickel-Aluminum).

Name of Engine:	**BLISS TITAN.**
Manufactured by:	E. W. Bliss Company, Brooklyn, N. Y.
Type:	5 cylinder radial, air cooled.
Rating:	225 H.P. at 1800 R.P.M.
Displacement:	842 cu. in.
Compression Ratio:	5 to 1.
Dimensions:	Bore .... 5¾" Stroke .... 6½"
Weight:	500 lbs. (dry).
Fuel Consumption at Rated H.P.:	Not more than .065 gal. per H.P. hr.
Oil Consumption:	Not more than .0012 gal. per H.P. hr.
Lubrication:	Pressure pumps.
Ignition:	2 H. T. Magnetos.
Carburetion:	Bristol Triplex.
Spark Plugs:	1 per cyl. K.L.G.
Price:	On application.

Accessories: (Cost Extra)
Eclipse Starter.

Special Features:

Major parts interchangeable with the Jupiter (9 cylinder) engine.

Low upkeep cost.

Accessories grouped at rear of engine for protection and convenience.

Name of Engine:	**CENTURY FOUR-IN-LINE.**
Manufactured by:	Century Rotary Motor Corporation, Canastota, New York.
Type:	4 cylinder, air cooled, vertical-in-line.
Rating:	100 H.P. at 1950 R.P.M.
Displacement:	310 cu. in.
Compression Ratio:	5.3 to 1.
Dimensions:	Length overall, 54 in. with supercharger Height, overall .... 31⅛″ Bore .... 4½″ Stroke .... 5″
Weight:	290 lbs.
Fuel Consumption at Rated H.P.:	With supercharger and atomizer, .356 lbs. per H.P. hr.; with Stromberg Carburetor N.A.R. 3, .500 lbs. per H.P. hr.
Oil Consumption:	Not more than .010 lbs. per H.P. hr.
Lubrication:	Pressure (gear pump). No scavenger pump.
Ignition:	2 Scintilla Magnetos.
Spark Plugs:	2 per cyl.
Price:	On application.

Equipment:

Electric self-starter with generator standard equipment.

Ignition Switch.

Instruction Pamphlet.

Special Features:

Nickel Cast Iron cylinders. Detachable nickel cast iron cylinder head with one intake and one exhaust valve.

Nickel Cast Iron Pistons.

Name of Engine:	**COMET 7-D.**
Manufactured by:	Comet Engine Corp., Madison, Wis.
Type:	7 cylinder, fixed radial air cooled. Approved Type Certificate No. 9.
Military Rating:	130 H.P. at 1825 R.P.M.
Commercial Rating:	165 H.P. at 1900 R.P.M.
Displacement:	611 cu. in.
Compression Ratio:	5 to 1.
Dimensions:	Length, overall .... 35⅝" Diameter, overall .... 46⅞" Bore .... 4½" Stroke .... 5½"
Weight:	395 lbs. (without starter or hub).
Fuel Consumption at Rated H.P.:	Not more than .55 lbs. per H.P. hr.
Oil Consumption:	Not more than .015 lbs. per H.P. hr.
Lubrication:	Force feed, scavenger and pressure pumps.
Ignition:	Dual Scintilla Magnetos.
Carburetion:	Stromberg.
Spark Plugs:	2 per cyl. AC.
Price:	$2,700.

Equipment:

Air Cleaner and Heater.
Nose Cowling.
Exhaust Manifold, complete.
Tool Kit.
Instruction Book.

Accessories: (Cost Extra)

Heywood Starter.

Special Features:

Valves placed fore and aft, with exhaust valves forward for better cooling, thus giving more uniform cylinder head temperatures.

One rocker arm, ball bearing mounted, operates both valves on each cylinder. Only one push rod per cylinder is used.

A single cam provides positive operation in both directions—push and pull—for all valve rods. The intake valves are operated by a downward pull on the valve rod. As a result, the tappet clearances on the intake valves decrease as the cylinders warm up, giving a longer intake opening at the higher speeds.

Due to positive operation of valve rods in both directions, valve spring tensions have been reduced materially, resulting in a very light and evenly distributed load on the entire valve mechanism.

Spark plugs being placed on the sides of the cylinder head give uniform spark plug temperatures.

Cylinders are composed of a steel barrel with integral mounting flange, over which a cast aluminum (Y-alloy) head is screwed and shrunk.

Crankshaft is carried in roller bearings, and is of two pieces for use of a solid master rod.

The crankcase is built of two sections united by seven through bolts, as well as by the cylinder flanges. The gear case cover carries the magneto and tachometer gears and the oil pumps. The diffuser cover supports the magnetos and carburetor. A ball thrust bearing is carried in the nose, or front section.

Number of gears in the cam and gear assembly is reduced to a minimum. Cam is ball bearing mounted.

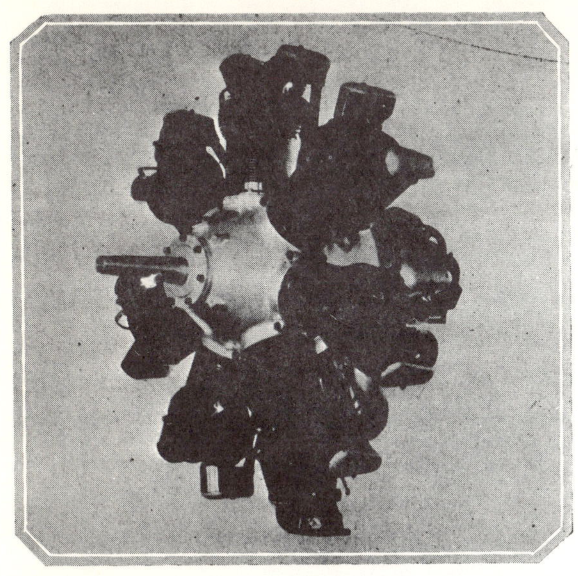

Name of Engine:	**CONTINENTAL A-70.**
Manufactured by:	Continental Aircraft Engine Co., Detroit, Mich.
Type:	7 cylinder, radial, air cooled.
Rating:	(Dept. of Commerce Approved Certificate), 165 H.P. at 2000 R.P.M.
Displacement:	544 cu. in.
Compression Ratio:	5.4 to 1.
Dimensions:	Length, overall ... 39" Diameter, overall ... 41¾" Bore ... 4⅝" Stroke ... 4⅝"
Weight:	430 lbs. without hub or starter.
Fuel Consumption at Rated H.P.:	Not more than .55 lbs. per H.P. hr.
Oil Consumption:	Not more than .035 lbs. per H.P. hr.
Lubrication:	Pressure and scavenger gear pumps, 75 lbs. to the sq. in.
Ignition:	2 Scintilla Magnetos.
Carburetion:	Stromberg.
Spark Plugs:	2 per cyl. B. G.
Price:	$2,700.

Equipment:

Exhaust Nose Ring and Piping complete.
Nose Cowling.
Air Heater and Cleaner Assemblies.
Priming Pump and Fittings.
Ignition Switch.
Tool Kit.

Accessories: (Cost Extra)

Any Starter (including Heywood) using standard SAE mounting cylinders drilled for Heywood fittings.
Any Generator with SAE flange.
Any Fuel Pump with SAE flange.
Propeller Hub.

Special Features:

Cast aluminum crankcase in two sections.

Two-piece forged chrome-nickel steel crankshaft.

One-piece master connecting rod. All rods "H" section.

Unribbed cast aluminum alloy pistons.

Composite steel and aluminum cylinder assemblies.

Heads screwed and shrunk to steel barrels.

Tulip valves, Tungsten intake, CNS exhaust.

Crankshaft mounted on ball and roller bearings.

All accessory drives except pump cross shaft on plain bearings with full pressure lubrication.

Rocker arms on ball bearings.

No rotary induction system. Double branch manifold feeding horseshoe header. Results in easy starting and excellent distribution and prompt acceleration.

Lubrication system eliminates all internal and external piping.

Engine may be installed in mounting ring without removal of any accessories.

Name of Engine:	**CROSLEY.**
Manufactured by:	Crosley Aircraft Company, Cincinnati and Sharonville, Ohio.
Type:	4 cylinder-in-line inverted, air cooled.
Commercial Rating:	100 H.P. at 2100 R.P.M.
Displacement:	301 cu. in.
Compression Ratio:	5 to 1.
Dimensions:	Length, overall .... 45⅝″ Width, overall .... 20″ Bore .... 4¼″ Stroke .... 5 5/16″
Weight:	300 lbs.
Fuel Consumption at Rated H.P.:	Not more than .56 lbs. per H.P. hr.
Oil Consumption:	Not more than .01 lbs. per H.P. hr.
Lubrication:	Pressure (Gear Pump) to all engine parts. Three gear scavenging pump, one pump functioning on climb, the other on glide.
Ignition:	Bosch Dual Type FU4R.
Carburetion:	Holley.
Spark Plugs:	2 per cyl.
Price:	On application.

Special Features:

One piece nickel iron cast cylinders with underhead camshaft having all valve gears enclosed in a cast aluminum housing which may be removed without disturbing engine timing. Camshaft driven by a spur gear train. Gears made of nitralloy steel. Generator drive provided.

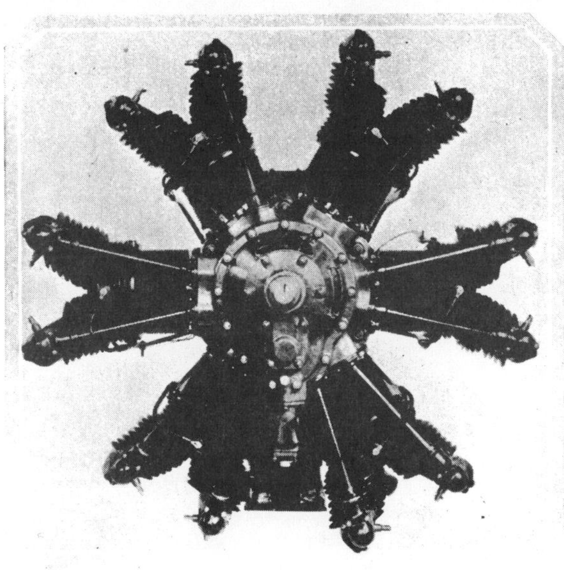

Name of Engine:	**CHALLENGER,** Model R-600.
Manufactured by:	Curtiss Aeroplane and Motor Co., Inc., Garden City and Buffalo, N. Y.
Type:	6 cylinder, Air Cooled, Fixed Radial (staggered)
Rating:	170 H.P. at 1800 R.P.M.
Displacement:	603 cu. in.
Compression Ratio:	Low compression engine 4.9 to 1. High compression engine 5.25 to 1.
Dimensions:	Length, overall ....42 5/32" Diameter, overall ....42⅝" Bore .... 5⅛" Stroke .... 4⅞"
Weight:	Dry, (without hub or starter)..420 lbs.
Fuel Consumption at Rated H.P.:	Not more than .55 lbs. per H.P. hr.
Oil Consumption:	Not more than .020 lbs. per H.P. hr.
Lubrication:	Pressure and scavenging pumps.
Ignition:	Two Scintilla.
Carburetion:	Stromberg 2 barrel, NA-U4-J.
Spark Plugs:	2 per cyl. B.G. 1XA
Price:	On application.

Accessories: (Cost Extra)

Propeller Hub.
Engine Driven Fuel Pump.
Starter.
Generator.
Gun Synchronizer.
Tool Kit.

Special Features:

Cast aluminum alloy heat treated cylinder heads screwed and shrunk onto forged steel barrels.

Rocker boxes cast integral with cylinder heads.

Crankcase cast in two parts, split on the center line of the front row of cylinders.

2 Silchrome valves per cylinder, seating on bronze inserts.

Two-throw counter balanced crankshaft, with two master rods each fitted with two short "H" section link rods.

Ribbed type aluminum alloy pistons.

Carburetor is exhaust jacketed around the barrels, using a hot-spot elbow fitted with a valve to control the amount of exhaust heat admitted to the carburetor.

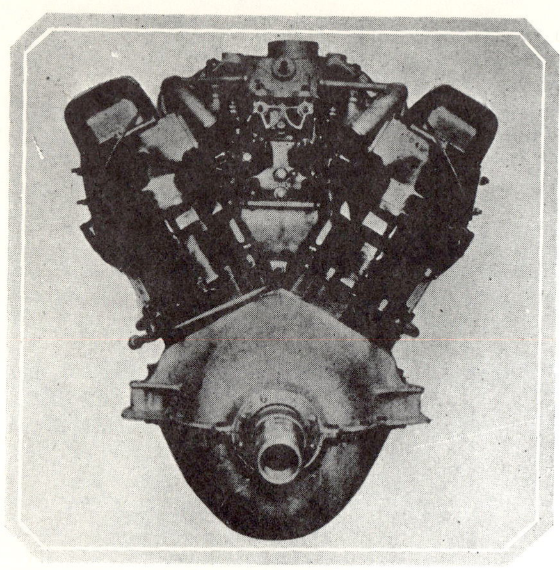

Name of Engine:	**CONQUEROR,** Models V-1570 and GV-1570
Manufactured by:	Curtiss Aeroplane and Motor Co., Inc., Garden City and Buffalo, N. Y.
Type:	12-cylinder, water-cooled, "V"-60° Direct Drive and Geared.
Rating:	600 H.P. at 2,400 R.P.M.—Direct. 600 H.P. at 2,450 R.P.M.—Geared.
Displacement:	1,569 cu. in.
Compression Ratio:	5.8 to 1.
Dimensions:	Height ....36 5/8" Length, overall.....63 1/16" direct 68 27/32" geared Width, overall......26 5/16" Bore .... 5 1/8" Stroke ....6 1/4"
Weight:	Dry, (without hub or starter), 755 lbs. direct 845 lbs. geared
Fuel Consumption at Rated H.P.:	Not more than .53 lbs. per H.P. hr.
Oil Consumption:	Not more than .015 lbs. per H.P. hr.
Lubrication:	Pressure and scavenging pumps.
Ignition:	Scintilla, dual.
Carburetion:	2 Stromberg Dual type NA-Y6-O.
Spark Plugs:	2 per cylinder B.G.
Price:	On application

Accessories: (Cost Extra)

Eclipse Hand Inertia Starter

Special Features:

Enbloc type jacket construction, in 2 banks of 6 each.
Wet sleeve type cylinders of heat treated hydraulically forged carbon steel screwed and shrunk into cylinder heads.
Four interchangeable tulip type valves seating directly in bronze inserts.
8 Bearing type crankshaft.
Gear case assembly is removable as a unit.
Ribbed head type aluminum alloy pistons.
Master and articulated type connecting rods.

Name of Engine:	**CURTISS D-12**
Manufactured by:	Curtiss Aeroplane and Motor Co., Inc., Garden City and Buffalo, N. Y.
Type:	12 cylinder, water cooled, "V"-60° Direct Drive.
Rating:	435 H.P. at 2300 R.P.M.
Displacement:	1145 cu. in.
Compression Ratio:	5.3 to 1.
Dimensions:	Height 34¾"
	Length, overall 56¾"
	Width, overall 28¼"
	Bore 4½"
	Stroke 6"
Weight:	Dry, (without special equipment) 680 lbs.
Fuel Consumption at Rated H.P.:	Not more than .53 lbs. per H.P. hr.
Oil Consumption:	Not more than .015 lbs. per H.P. hr.
Lubrication:	Pressure and scavenging pumps.
Ignition:	2 Scintilla.
Carburetion:	2 Stromberg Dual type NA-Y5-F.
Spark Plugs:	2 per cylinder B.G.
Price:	On application.

Accessories: (Cost Extra)
Eclipse Hand Inertia Starter.

Special Features:

Enbloc type jacket construction in 2 banks of 6 each.
Cylinders of heat treated hydraulically forged carbon steel with head end closed.
Four interchangeable tulip type valves seating directly in each cylinder head.
8 Bearing type crankshaft.
Gearcase assembly is removable as a unit.
Ribbed head type aluminum alloy pistons.
Master and articulated type connecting rods.

Name of Engine:	**DAYTON BEAR.**
Manufactured by:	Dayton Airplane Engine Co., Dayton, Ohio.
Type:	4-in-line, air cooled (Approved Type Certificate No 11).
Military Rating:	110 H.P at 1500 R.P.M.
Commercial Rating:	120 H.P. at 1850 R.P.M
Displacement:	445 cu. in.
Compression Ratio:	5.1 to 1.
Dimensions:	Length, overall ....50 3/32" Height, overall ....38 27/32" Width ....18¼" Bore .... 4½" Stroke .... 7"
Weight:	375 lbs.
Fuel Consumption at Rated H.P.:	Not more than .525 lbs per H.P. hr.
Oil Consumption:	Not more than .0125 lbs. per H.P. hr.
Lubrication:	Force feed.
Ignition:	2 Dual Bosch Magnetos.
Carburetion:	Zenith L-8.
Spark Plugs:	2 per cyl. B. G. 1 XA.
Price:	$1,450.

Equipment:

Instruction Book.
Tool Kit.

Hub for Wood Propeller.

Exhaust Flanges.

Accessories: (Cost Extra)

Fuel Pump Drive.
C-5 Fuel Pump.

Starter.

Hub for Steel Propeller.

Special Features:

Cast Nickel Iron cylinder with detachable aluminum alloy head.

Bronze valve seats shrunk in head.

One exhaust valve and one intake valve, special tulip valve design of special steel.

Aluminum alloy pistons.

Full floating hollow piston pin.

1-piece crankshaft—5 bearings.

Hot spot effect secured by passing oil around intake manifold.

Name of Engine:	DAYTON GRANT.
Manufactured by:	Dayton Airplane Engine Co., Dayton, Ohio.
Type:	8 cylinder horizontal opposed.
Military Rating:	225 H.P at 1800 R.P.M.
Commercial Rating:	240 H.P. at 2000 R.P.M.
Displacement:	764 cu. in.
Compression Ratio:	5.3 to 1.
Dimensions:	Length, overall ... 56" Width, overall ... 44" Height ... 20" Bore ... 4½" Stroke ... 6"
Weight:	500 lbs.
Fuel Consumption at Rated H.P.:	Not more than 1.05 lbs. per H.P. hr.
Oil Consumption:	Not more than .025 lbs. per H.P. hr.
Lubrication:	Force feed.
Ignition:	Dual spark magneto or generator for battery ignition.
Carburetion:	2 Zenith or Stromberg.
Spark Plugs:	2 B. G. per cyl. 1 XA.
Price:	On Application.

Equipment:

Instruction Book. Tool Kit. Exhaust Flanges.

Accessories: (Cost Extra)

C-5 Fuel Pump.
Hub for Wood Propeller.
Starter.
Hub for Steel Propeller.

Special Features:

Steel cylinder with detachable aluminum alloy head.

Bronze valve seats shrunk in head.

One exhause valve and one intake valve, special tulip valve design of special steel.

Aluminum alloy pistons.

Full floating hollow piston pin.

1-piece crankshaft, 5 bearings.

Side by side connecting rods.

Name of Engine:	**GENET.**
Manufactured by:	Fairchild Airplane Mfg. Co., Farmingdale, N. Y.
Type:	5 cylinder radial, air cooled (Approved Type Certificate).
Rating:	88 H.P. at 2200 R.P.M.
Displacement:	251 cu. in.
Compression Ratio:	5.2 to 1.
Dimensions:	Length, overall .... 28½″ Diameter, overall .... 36″ Bore .... 4″ Stroke .... 4″
Weight:	200 lbs.
Fuel Consumption at Rated H.P.:	Not more than .55 lbs. per H.P. hr.
Oil Consumption:	Not more than .025 lbs. per H.P. hr.
Lubrication:	Dry sump, pressure and scavenger pumps.
Ignition:	Dual B. T. H.
Carburetion:	Claudel.
Spark Plugs:	2 per cyl., B.G.
Price:	$1,500.

Accessories: (Cost Extra)
Starter.

Special Features:

Left hand tractor type.

Cylinders—Steel forged, screwed into crankcase adaptors.

Cylinders Heads—Aluminum alloy castings screwed, shrunk and locked to cylinders. Valve seats and plug bosses cast in position.

Connecting Rods—"H" section.

Pistons—"Y" alloy fitted with 2 compression and 2 scraper rings.

Overhead valves and gear.

Duplex valve springs.

Name of Engine:	**MOORE THREE VALVE.**
Manufactured by:	General Airmotors Co., Inc. Scranton, Pa.
Type:	5 cylinder fixed radial, air cooled.
Commercial Rating:	150 H.P. at 1850 R.P.M.
Displacement:	540 cu. in.
Compression Ratio:	5.4 to 1.
Dimensions:	Length, overall ... 41½″ Diameter, overall ... 44⅝″ Bore ... 5″ Stroke ... 5½″
Weight:	365 lbs. (net).
Fuel Consumption at Rated H.P.:	Not more than .53 lbs. per H.P. hr.
Oil Consumption:	Not more than .015 lbs. average per H.P. hr.
Lubrication:	Force feed.
Ignition:	2 Scintilla Magnetos PN-5-D.
Carburetion:	1 Stromberg NA R 5A.
Spark Plugs:	2 per cyl.
Price:	On application.

Accessories:

Heywood Starter. Propeller Hub. Nose Cowling and Exhaust Manifold.

Special Features:

Compression control mechanism for easy altitude adjustment.

Special valve arrangement—3 valves per cylinder, 2 intake, 1 exhaust giving maximum valve area.

Poultice head secured to cylinder barrel by patented conforming clamp.

Valve rocker arm assembly is a unit of three parts and three sets of ball bearings. Valves seat in steel cylinder.

Low head temperature, maximum 350 degrees F. Oil temperature, 120 degrees F.

Crankcase assembly, two major Bohnalite castings.

Two piece crankshaft, counterbalanced. Solid master rod.

Full floating main bearing on crank pin.

Cylinders, carbon steel cup shaped blanks, forged and bored.

Name of Engine:	**MOORE THREE VALVE.**
Manufactured by:	General Airmotors Co., Inc., Scranton, Pa.
Type:	7 cylinder fixed radial, air cooled.
Commercial Rating:	210 H.P. at 1850 R.P.M.
Displacement:	756 cu. in.
Compression Ratio:	5.4 to 1.
Dimensions:	Length, overall ..... 41½" Diameter, overall ..... 44⅝" Bore ..... 5" Stroke ..... 5½"
Weight:	425 lbs. (net)
Fuel Consumption at Rated H.P.:	Not more than .53 lbs. per H.P. hr.
Oil Consumption:	Not more than .015 lbs. average per H.P. hr.
Lubrication:	Force feed.
Ignition:	2 Scintilla Magnetos PN-5-D.
Carburetion:	1 Stromberg NA R 5A.
Spark Plugs:	2 per cyl.
Price:	On application.

Accessories:

Heywood Starter. Propeller Hub. Nose Cowling and Exhaust Manifold.

Special Features:

Compression control mechanism for easy altitude adjustment.

Special valve arrangement—3 valves per cylinder, 2 intake, 1 exhaust giving maximum valve area.

Poultice head secured to cylinder barrel by patented conforming clamp.

Valve rocker arm assembly is a unit of three parts and three sets of ball bearings. Valve seat in steel cylinder.

Low head temperature, maximum 350 degrees F. Oil temperature, 120 degrees F.

Crank case assembly, two major Bohnalite castings.

Two-piece crankshaft, counterbalanced. Solid master rod. Full floating main bearing on crank pin.

Cylinders, carbon steel cup shaped blanks, forged and bored.

Name of Engine:	**IRWIN 79**
Manufactured by:	Irwin Aircraft Co., Sacramento, Calif.
Type:	4 cylinder, radial, air cooled.
Commercial Rating:	20 H.P. at 1730 R.P.M.
Displacement:	79 cu. in.
Compression Ratio:	4.9 to 1.
Dimensions:	Length, overall .... 18″ Diameter, overall .... 23″ Bore .... 2⅞″ Stroke .... 2¾″
Weight:	58 lbs.
Fuel Consumption at Rated H.P.:	Not more than .50 lbs. per H.P. hr.
Oil Consumption:	Not more than .06 lbs. per H.P. hr.
Lubrication:	Jet spray cylinder walls and pressure to main bearings.
Ignition:	1 Bosch.
Carburetion:	1 Winfield.
Spark Plugs:	1 per cyl. Robert Bosch.
Price:	$625.

Accessories: (Cost Extra)

Propeller.

Tachometer.

Special Features:

Cylinders are composed of a nickel cast iron barrel over which an aluminum alloy cylinder is shrunk and doweled. Valve seats are of cast iron shrunk and doweled in place.

The crankcase assembly is composed of two major castings of aluminum alloy.

The crankshaft is made from special heat treated chrome nickel steel, machined and ground from the solid billet weighing approximately 90 lbs., while the finished shaft weighs but 6½ lbs.

Alloy steel heat treated connecting rods of H section fitted with bronze gudgeon pin bushings.

Pistons of aluminum alloy, fitted with full floating nickel steel hollow gudgeon pins.

Main bearings are of the ball race type generously lubricated under pressure.

Name of Engine:	**BEETLE,** Model K.
Manufactured by:	Kimball Aircraft Corp., Naugatuck, Conn.
Type:	7 cylinder radial, air cooled.
Rating:	135 H.P. at 1800 R.P.M. (Approved Type Certificate No. 34).
Displacement:	585 cu. in.
Compression Ratio:	5.2 to 1.
Dimensions:	Length, overall .... 31¾″ Diameter, overall .... 45″ Bore .... 4½″ Stroke .... 5¼″
Weight:	370 lbs.
Fuel Consumption at Rated H.P.:	Not more than .54 lbs. per H.P. hr.
Oil Consumption:	Not more than .02 lbs. per H.P. hr.
Lubrication:	Dry sump, force feed and splash.
Ignition:	Dual Scintilla.
Carburetion:	1 Stromberg NA-S5.
Spark Plugs:	2 per cyl. AC.
Price:	$2,900.

Accessories: (Cost Extra)
Any type starter.

Special Features:

Removable exhaust valve cage.
Ball bearing cam ring.
Universal mounting ring.

Name of Engine:	**KINNER K-5.**
Manufactured by:	Kinner Airplane & Motor Corporation, 635 West Colorado Boulevard, Glendale, California.
Type:	5 cylinder, radial, air cooled, 4 cycle. (Approved Type Certificate No. 3.)
Department of Commerce Rating:	90 H.P. at 1810 R.P.M.
Commercial Rating:	110 H.P. at 1880 R.P.M.
Displacement:	372 cu. in.
Compression Ratio:	5.0 to 1.
Dimensions:	Length, overall ... 32¼" Diameter, overall ... 43" Bore ... 4¼" Stroke ... 5¼"
Weight:	278 lbs. (without starter or hub).
Fuel Consumption at Rated H.P.:	Not more than .60 lbs. per H.P. hr.
Oil Consumption:	Not more than .025 lbs. per H.P. hr.
Lubrication:	Circulating oil system, pressure feed to front and rear main bearings and to link pins which are all of the plain type, through grooved master rod bearing.
Ignition:	Dual Scintilla.
Carburetion:	1 Stromberg NAR-5-A or Holley.
Spark Plugs:	2 per cyl. B. G. No. 4 Hornet.
Price:	$1,800 f.o.b. Glendale.

Accessories: (Cost Extra)

Heywood Self-starter. Eclipse Hand Turning Gear. Propeller Hub.

Special Features:

A light, simple 100 H.P. motor delivering a degree of dependable performance equal in every respect to that obtained from the powerful military motors in the 400-500 H.P. class.

Crankcase is manufactured from heat treated aluminum alloy castings, consisting of three pieces, the main cylinder sections and front and rear covers. The removal of the covers makes immediately accessible, the crankshaft, bearings, master and link rod, cam gears, followers and magneto drive gears.

The crankshaft is of a special heat-treated chrome nickel steel drop-forging, machined all over, counter balanced and supported on two main bearings of special bronze backed babbit. The thrust bearing is a Hoffman ball bearing.

Special force feed lubrication to the link pins.

A special air heater is provided permitting air to be drawn from top or side of fuselage.

Special annular intake passage cast in main crankcase giving excellent distribution of combustible mixture.

Valve rocker arm enclosed in streamlined cover.

Inlet and exhaust valves are both necked type of special heat resisting steel alloy.

Individual camshafts for each cylinder.

Name of Engine:	**KINNER R-715.**
Manufactured by:	Kinner Airplane & Motor Corporation, 635 West Colorado Boulevard, Glendale, California.
Type:	5 cylinder, radial, air cooled, 4 cycle.
Manufacturer's Rating:	190 H.P. at 1800 R.P.M.
Displacement:	715 cu. in.
Compression Ratio:	5 to 1.
Dimensions:	Diameter, overall ....50″ Length, overall ....34 13/16″ Bore .... 5⅝″ Stroke ....5¾″
Weight:	410 lbs., dry, without hub, carburetor air heater, exhaust collector ring, starter, generator, fuel pump.
Lubrication:	Circulating oil system, pressure feed to front and rear main bearings and to link pins which are all of the plain type, through grooved master rod bearing.
Ignition:	Two Scintilla Magnetos.
Carburetion:	Stromberg.
Price:	On application.

Special Features:

Outstanding features of the 715 cubic inch model are simplicity, smoothness of operation, light weight for its power and excellent forward visibility. The lack of interference with airflow at the nose of the airplane (tractor type) insures good aerodynamic properties (no interference with empenage).

This engine can be used either as tractor or pusher.

Standard starter mount affords a wide range of starters, as Heywood injection, hand turning gear, inertia hand, electric or combination starters, electric starter, etc. The provision of standard drive for an electric starter affords use of electric type starters, lighting system, radio, etc.

Provision of fuel pump drive permits considerable latitude in location of fuel tanks. However, the carburetor location is such that a simple gravity system can be used.

Standard S. A. E. propeller end of crankshaft provides for use of minimum weight of metal propeller hub.

Accessories are standard proven types.

Name of Engine:	**THE MAJOR LAMBERT,** Model R-266.
Manufactured by:	Lambert Aircraft Engine Corp., Moline, Ill.
Type:	5 cylinder, fixed radial, air cooled.
Department of Commerce Rating:	90 H.P. at 2375 R.P.M.
Displacement:	266 cu. in.
Compression Ratio:	5.55 to 1.
Dimensions:	Length, overall .... 30⅜″ Diameter, overall .... 33″ Bore .... 4¼″ Stroke .... 3¾″
Weight:	(Minus equipment) 214 lbs.
Fuel Consumption at Rated H.P.:	Not more than .55 lbs. per H.P. hr.
Oil Consumption:	Not more than .025 lbs. per H.P. hr.
Lubrication:	Pressure and Dry Sump.
Ignition:	Scintilla Dual.
Carburetion:	Stromberg NAR 3.
Spark Plugs:	2 per cyl. (Champion).
Price:	On application.

Equipment:

Propeller Hub.
Exhaust Ring.
Carburetor Air Heater and Muffler.
Tool Kit.
Instruction and Parts Book.

Special Features:

Small overall diameter and clean lines, ease of assembly, low operating and maintenance costs, and ruggedness combined with unusually light weight for an engine of its power class.

Solid master connecting rod and two-piece crankshaft which makes high crank speeds possible.

Two-piece magnesium alloy crankcase, split on the cylinder centerline and held together with ten bolts. The front section carries the front main ball-bearing and thrust ball-bearing. The inlet passage and valve lifter guides are cast in the rear section.

Cylinder barrels are nickel semi-steel castings.

Removable cylinder heads of aluminum alloy with exhaust and inlet ports toward the rear, are held in place with four long studs which run through to the crankcase.

Enclosed valve actuating mechanism is at the rear of the cylinders.

Accessories and their drives are grouped at the rear making for easy cowling.

Name of Engine:	**LE BLOND "60,"** Model 5-D.
Manufactured by:	Le Blond Aircraft Engine Corp., Cincinnati, Ohio.
Type:	5 cylinder, radial, air cooled.
Rating:	65 H.P. at 1950 R.P.M. (Approved Type Certificate No. 12).
Displacement:	351 cu. in.
Compression Ratio:	5.42 to 1.
Dimensions:	Length, overall 22¼" Diameter, overall 32¾" Bore 4⅛" Stroke 3¾"
Weight:	222 lbs. dry complete.
Fuel Consumption at Rated H.P.:	Not more than .55 lbs. per H.P. hr.
Oil Consumption:	Not more than .015 lbs. per H.P. hr.
Lubrication:	Dry sump, full pressure feed.
Ignition:	Dual Robert Bosch.
Carburetion:	1 Stromberg.
Spark Plugs:	2 per cyl. Champion.
Price:	$1,230.

Accessories: (Cost Extra)
Eclipse Starter.

Name of Engine:	**LE BLOND "90,"** Model 7-D.
Manufactured by:	Le Blond Aircraft Engine Co., Cincinnati, Ohio.
Type:	7 cylinder radial, air cooled.
Rating:	90 H.P. at 1975 R.P.M. (Approved Type Certificate No. 20.)
Displacement:	351 cu. in.
Compression Ratio:	5.42 to 1.
Dimensions:	Length, overall .... 23 1/16" Diameter, overall .... 32¾" Bore .... 4⅛" Stroke .... 3¾"
Weight:	285 lbs. dry complete.
Fuel Consumption at Rated H.P.:	Not more than .55 lbs. per H.P. hr.
Oil Consumption:	Not more than .015 lbs. per H.P. hr.
Lubrication:	Dry sump, full pressure feed.
Ignition:	Dual Scintilla.
Carburetion:	Stromberg.
Spark Plugs:	2 per cyl. Champion.
Price:	$1,610.

Accessories: (Cost Extra)
Eclipse Electric Starter.

Name of Engine:	**BROWNBACK TIGER C-400.**
Manufactured by:	Light Manufacturing & Foundry Co., Pottstown, Pa.
Type:	6 cylinder, double row radial, air cooled; 4 strokes per cycle.
Rating:	90 H.P. at 1700 R.P.M. Dept. of Commerce Certificate No. 33.
Displacement:	397 cu. in.
Compression Ratio:	4.8 to 1.
Dimensions:	Length, overall ....29 3/16″ Diameter, overall ....37″ Bore ....4.13″ Stroke ....4.93″
Weight:	275 lbs., less hub and starter.
Fuel Consumption at Rated H.P.:	Not more than .56 lbs. per H.P. hr.
Oil Consumption:	Not more than .011 lbs. per H.P. hr.
Lubrication:	Dry sump; force feed.
Ignition:	2 Bosch Magnetos.
Carburetion:	1 Stromberg.
Spark Plugs:	2 per cyl. Champion.
Price:	$1,840.

Accessories: (Cost Extra)

Any 5″ Standard Starter.

Exhaust Manifold Stacks optional.

Special Features:

Overhead valves through rockers and push rods.

Aluminum alloy cylinder head, cast iron barrel.

One Silchrome inlet valve per cylinder.

One Silchrome exhaust valve per cylinder.

Single-piece crankshaft of nickel-chrome steel with double throw design giving perfect balance with a minimum of counterbalance weight.

Name of Engine:	**LYCOMING R-680.**
Manufactured by:	Lycoming Manufacturing Co., 652 Oliver St., Williamsport, Pa.
Type:	9-cylinder, radial, air-cooled.
Rating:	210 H.P. at 2000 R.P.M.
Displacement:	680 cu. in.
Compression Ratio:	5.2 to 1.
Dimensions:	Length, overall (Including propeller hub and starter) ...... 43⅛" Diameter, overall .... 43¼" Bore .... 4⅝" Stroke .... 4½"
Weight:	Dry, 465 lbs.
Fuel Consumption at Rated H.P.:	Not more than .55 lbs. per H.P. hr.
Oil Consumption:	Not more than .035 lbs. per H.P. hr.
Lubrication:	1 Pressure Pump. 2 Scavenger Pumps (in one unit). Pressure lubrication, to crankshaft, master rod, link pins, cam drive gears, and main accessories drive shaft and gears.
Ignition:	Scintilla Dual Magneto Type SC-A.
Carburetion:	1 Stromberg NA-R7.
Spark Plugs:	2 per cyl.
Price:	On application.

Equipment:

Tachometer Drive—½ Crankshaft.
Air Cleaner.
Exhaust Ring.
Cowling.

Accessories:

Generator.
Starter.

Name of Engine:	**PANTHER.**
Manufactured by:	MacClatchie Manufacturing Co., Box 189, Compton, Calif.
Type:	7 cylinder, radial, L-head, air cooled, 4 cycle.
Rating:	150 H.P. at 1900 R.P.M.
Displacement:	612.15 cu. in.
Compression Ratio:	5 to 1.
Dimensions:	Length, overall ....35″ Diameter, overall ....36″ Bore .... 4½″ Stroke .... 5½″
Weight:	400 lbs.
Fuel Consumption at Rated H.P.:	Not more than .56 lbs. per H.P. hr.
Oil Consumption:	Not more than .023 lbs. per H.P. hr.
Lubrication:	Pressure system.
Ignition:	Dual Scintilla.
Carburetion:	Stromberg NAR-5A.
Spark Plugs:	2 per cyl.
Price:	On request.

Special Features:

L-Head—"L" shaped turbulent combustion chambers in the cylinder heads.

Rocker arms, long pushrods and all allied parts are eliminated. This is made possible through the L-head type of construction, adapted successfully to a radial aviation engine for the first time in this power plant. The valves are actuated by direct contact with the cam rather than by rocker arms working on long pushrods.

Diameter only 36 inches. This small frontal elevation considerably decreases the wind resistance, making for better stream-lining of the unit.

Fewer Parts—Engineers estimate that this L-head radial engine has some 200 less parts than other power plants in its horsepower class.

Valves can be ground without removing the cylinders. The port plugs are unscrewed, and the valves may be removed quickly through the ports.

Name of Engine:	**ROVER.**
Manufactured by:	Michigan Aero-Engine Corporation, Lansing, Mich.
Type:	4 cylinder-in-line, inverted, air cooled.
Commercial Rating:	75 H.P. at 1975 R.P.M.
Displacement:	267.28 cu. in.
Compression Ratio:	5.15 to 1.
Dimensions:	Length, overall, including hub and mag. controls .... 42¼″
	Width, overall .... 11⅛″
	Height, overall .... 31⅞″
	Bore .... 4⅛″
	Stroke .... 5″
Weight:	232 lbs., less starter.
Fuel Consumption at Rated H.P.:	Not more than .18 lbs. per H.P. hr.
Oil Consumption:	Not more than .055 lbs. per H.P. hr.
Lubrication:	3 gear scavenging pump. 2 gear pressure pump. Removable oil filtering screen.
Ignition:	2 Scintilla Magnetos.
Carburetion:	1 Stromberg NA*R3 with altitude adjustment.
Spark Plugs:	2 per cyl. AC.
Price:	$1,200.00

Equipment:

Main Cooling Air Scoop.
Complete Exhaust System.
Carburetor Air Heater with control.
Propeller Hub.
Tool Kit.
Instruction Book.

Accessories: (Cost Extra)

Starter (hand or electric).

Special Features:

Aluminum alloy cylinder heads.

Cast Nickel Iron cylinders.

Completely enclosed valve gear ports.

Inverted type, affording many interesting installation features.

Compact arrangement with all accessories arranged at rear.
Integral, 5 bearing crankshaft.

Single, 5 bearing crankshaft in crankcase.

Ribbed, die-cast aluminum alloy pistons.

Dropped forged "Dural" connecting rods.

Lubrication system designed completely within the engine.

Dry sump, all oil carried in an external tank.

Independent scavenger system for each end of crankcase.

Name of Engine:	**PACKARD A-1500.**
Manufactured by:	Packard Motor Car Co., Detroit, Mich.
Type:	12 cylinder V-type water cooled. (Approved Type Certificate No. 18.)
Military Rating:	525 H.P. at 1900 R.P.M.
Commercial Rating:	600 H.P. at 2500 R.P.M.
Displacement:	1530 cu. in.
Compression Ratio:	5.1 to 1.
Dimensions:	Length, overall ....65 1/16″ Width, overall ....27″ Height ....39 Bore .... 5⅜″ Stroke .... 5½″
Weight:	800 lbs.
Fuel Consumption at Rated H.P.:	Not more than .55 lbs. per H.P. hr.
Oil Consumption:	Not more than .025 lbs. per H.P. hr.
Lubrication:	Pressure, dry sump, double scavenged.
Ignition:	Delco or Scintilla dual.
Carburetion:	Dual Stromberg NAY 6P.
Spark Plugs:	2 per cyl. B. G. Hornet.
Price:	$12,000 (direct drive). $17,375 (geared).

Equipment:

Fuel Pump. Propeller Hub attaching parts. Tool Kit.

Name of Engine:	**PACKARD A-2500.**
Manufactured by:	Packard Motor Car Co., Detroit, Mich.
Type:	12 cylinder, V-type, water-cooled. (Approved Type Certificate No. 19.)
Military Rating:	835 H.P. at 2100 R.P.M.
Commercial Rating:	800 H.P. at 2000 R.P.M.
Displacement:	2,540 cu. in.
Compression Ratio:	5.1 to 1.
Dimensions:	Length, overall .... 69¾" Width, overall .... 29½" Height .... 42" Bore .... 6⅜" Stroke .... 6½"
Weight:	1210 lbs
Fuel Consumption at Rated H.P.:	Not more than .52 lbs. per H.P. hr.
Oil Consumption:	Not more than .025 lbs. per H.P. hr.
Ignition:	Delco or Scintilla dual.
Carburetion:	Dual Stromberg NAS-12
Spark Plugs:	2 per cyl. B. G. Hornet.
Price:	$15,625 (direct drive). $19,375 (geared).

Equipment:

Fuel Pump. Propeller Hub attaching parts. Tool Kit

Accessories: (Cost Extra)

Aeromarine Starter.

Name of Engine:	**PACKARD X-24.**
Manufactured by:	Packard Motor Car Co., Detroit, Mich.
Type:	24-cylinder, x-type, water-cooled.
Military Rating:	1200 H.P. at 2600 R.P.M.
Commercial Rating:	1200 H.P. at 2600 R.P.M.
Displacement:	2775 cu. in.
Compression Ratio:	6 to 1.
Dimensions:	Length, overall ....77½" Height ....44 33/64" Width ....28 5/16" Bore ....5⅜" Stroke ....5"
Fuel Consumption at Rated H.P.:	Not more than .55 lbs. per H.P. hr.
Oil Consumption:	Not more than .035 lbs. per H.P. hr.
Lubrication:	Pressure, dry sump, double scavenged.
Ignition:	4 Delco.
Carburetion:	4 NAY6 P—Stromberg.
Spark Plugs:	2 Hornet B.G. per cyl.
Price:	On application.

Equipment:

Tool Kit. Fuel Pump. Propeller Hub and attaching parts.

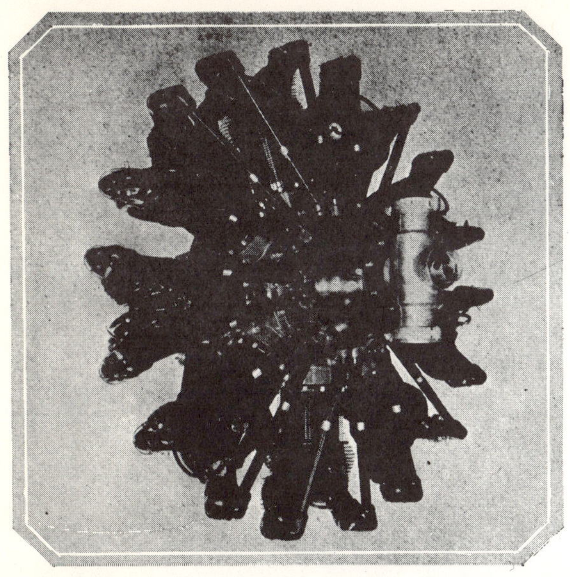

Name of Engine:	**HORNET,** Series A-1.
Manufactured by:	The Pratt & Whitney Aircraft Co., Hartford, Conn.
Type:	9 cylinder, fixed radial, air cooled.
Military Rating:	525 H.P. at 1900 R.P.M.
Commercial Rating:	525 H.P. at 1900 R.P.M. direct. 500 H.P. at 1900 R.P.M. geared.
Displacement:	1690 cu. in.
Compression Ratio:	5 to 1.
Dimensions:	Length, overall ....44⅞" Diameter, overall ....55-7/16" Bore ....6.125" Stroke ....6.375"
Weight:	Without special equipment, 775 lbs.
Fuel Consumption at Rated H.P.:	Not more than .55 pounds per H.P. hr.
Oil Consumption:	Not more than .035 lbs. per H.P. hr.
Lubrication:	Pressure (Gear Pump) 75 to 100 lbs.
Ignition:	Scintilla, dual.
Carburetion:	Stromberg, 2 bbls.
Starter:	Eclipse.
Spark Plugs:	B.G.
Price:	Series A—$8,500. Series A.G.—$9,500.

Special Features:

Solid master connecting rod and two-piece crankshaft. The master connecting rod has a solid instead of detachable cap big end. This construction makes possible high crank speeds which have been impossible with the two-piece rod. The single throw two-piece crankshaft is divided into a forward and rear section. The crankpin is integral with the forward section which transmits power to the propeller hub carried by it. The rear section telescopes into the crankpin and is carried completely through it. The two sections are united by a through bolt and kept in the proper angular relation by splines.

Forged aluminum crank case. The main crankcase of forged aluminum is divided into two similar sections in the plane of the cylinder and united by nine through bolts between the cylinders as well as by the cylinder flanges.

Enclosed valve gear. All valve operating parts are enclosed. The rocker arms are supported by ball bearings and are mounted in the rocker housings which are part of the cylinder head. The push rods are enclosed by telescopic covers held in place by springs.

Built in Supercharger. Every Pratt & Whitney engine is provided with a General Electric rotary induction fan used to provide proper mixture distribution. With suitable gear ratio it is possible for certain specialized purposes to provide a reasonable amount of supercharging without additional weight or complication.

Grouping of accessories at the rear. All the accessories are grouped at the rear of the engine protected by the cowling from weather and provide simplicity in control hookup and engine cowling.

On geared type engines a 2:1 propeller speed reduction is provided by a geared unit of patented Pratt & Whitney design. (Weight 840 lbs.)

Name of Engine:	**HORNET,** Series B.
Manufactured by:	The Pratt & Whitney Aircraft Co., Hartford, Conn.
Type:	9 cylinder, fixed radial, aircooled.
Rating:	575 H.P. at 1950 R.P.M. direct. 550 H.P. at 1950 R.P.M. geared
Displacement:	1860 cubic inches.
Compression Ratio:	5:1.
Dimensions:	Length overall ....44 9/16" Diameter ....56¾" Bore .... 6.25 Stroke 6.75
Weight:	Without special equipment approx. 810 lbs.
Fuel Consumption at Rated H.P.:	Not more than .55 lbs./H.P./hr.
Oil Consumption:	Not more than .035 lbs./H.P./hr.
Lubrication:	Pressure (Gear Pump) 75 to 100 lbs.
Ignition:	Scintilla, dual.
Carburetion:	Stromberg, 2 bbls.
Starter:	Eclipse.
Spark Plugs:	B.G.
Price:	Series B—$8,500. Series B G—$9,500.

Special Features:

Solid master connecting rod and two-piece crankshaft. The master connecting rod has a solid instead of detachable cap big end. This construction makes possible high crank speeds which have been impossible with the two-piece rod. The single throw two-piece crankshaft is divided into a forward and rear section. The crankpin is integral with the forward section which transmits power to the propeller hub carried by it. The rear section telescopes into the crankpin and is carried completely through it. The two sections are united by a through bolt and kept in the proper angular relation by splines.

Forged aluminum crank case. The main crankcase of forged aluminum is divided into two similar sections in the plane of the cylinder and united by nine through bolts between the cylinders as well as by the cylinder flanges.

Enclosed valve gear. All valve operating parts are enclosed. The rocker arms are supported by ball bearings and are mounted in the rocker housings which are part of the cylinder head. The push rods are enclosed by telescopic covers held in place by springs.

Built-in Supercharger. Every Pratt & Whitney engine is provided with a General Electric rotary induction fan used to provide proper mixture distribution. With suitable gear ratio it is possible for certain specialized purposes to provide a reasonable amount of supercharging without additional weight or complication.

Grouping of accessories at the rear. All the accessories are grouped at the rear of the engine protected by the cowling from weather and provide simplicity in control hookup and engine cowling.

On geared type engines a 3:2 propeller speed reduction is provided by a geared unit of patented Pratt & Whitney design. (Weight 885 lbs.)

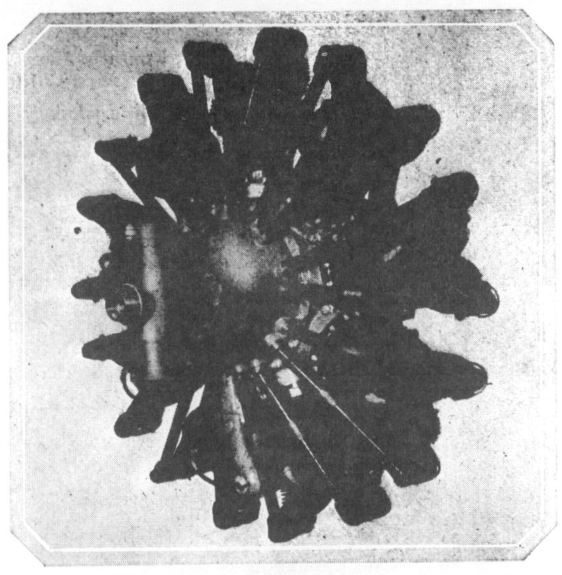

Name of Engine:	**WASP, Series C.**
Manufactured by:	The Pratt & Whitney Aircraft Co., Hartford, Conn.
Type:	9 Cylinder, fixed radial, air cooled.
Military Rating:	450 H.P. at 2100 R.P.M.
Commercial Rating:	420 H.P. at 2000 R.P.M. direct. 420 H.P. at 2100 R.P.M. geared.
Displacement:	1344 cu. in.
Compression Ratio:	5.25 to 1.
Dimensions:	Length, overall ....42-9/16" Diameter ....51½" Bore .... 5¾" Stroke .... 5¾"
Weight:	Dry, without special equipment, 695 lbs.
Fuel Consumption at Rated H.P.:	Not more than .55 lbs. per H.P. hr.
Oil Consumption:	Not more than .035 lbs. per H.P. hr.
Lubrication:	Pressure (Gear Pump) 75 to 100 lbs.
Ignition:	Scintilla, dual.
Carburetion:	Stromberg, 2 bbls.
Starter:	Eclipse.
Spark Plugs:	B.G.
Price:	Series C—$7,200. Series C.G.—$8,200.

Special Features:

Solid master connecting rod and two-piece crankshaft. The master connecting rod has a solid instead of detachable cap big end. This construction makes possible high crank speeds which have been impossible with the two-piece rod. The single throw two-piece crankshaft is divided into a forward and rear section. The crankpin is integral with the forward section which transmits power to the propeller hub carried by it. The rear section telescopes into the crankpin and is carried completely through it. The two sections are united by a through bolt and kept in the proper angular relation by splines.

Forged aluminum crankcase. The main crankcase of forged aluminum is divided into two similar sections in the plane of the cylinder and united by nine through bolts between the cylinders as well as by the cylinder flanges.

Enclosed valve gear. All valve operating parts are enclosed. The rocker arms are supported by ball bearings and are mounted in the rocker housings which are part of the cylinder head. The push rods are enclosed by telescopic covers held in place by springs.

Built in Supercharger. Every Pratt & Whitney engine is provided with a General Electric rotary induction fan used to provide proper mixture distribution. With suitable gear ratio it is possible for certain specialized purposes to provide a reasonable amount of supercharging without additional weight or complication.

Grouping of accessories at the rear. All the accessories are grouped at the rear of the engine protected by the cowling from weather and provide simplicity in control hookup and engine cowling.

On geared type engines a 2:1 propeller speed reduction is provided by a geared unit of patented Pratt & Whitney design. (Weight 775 lbs.)

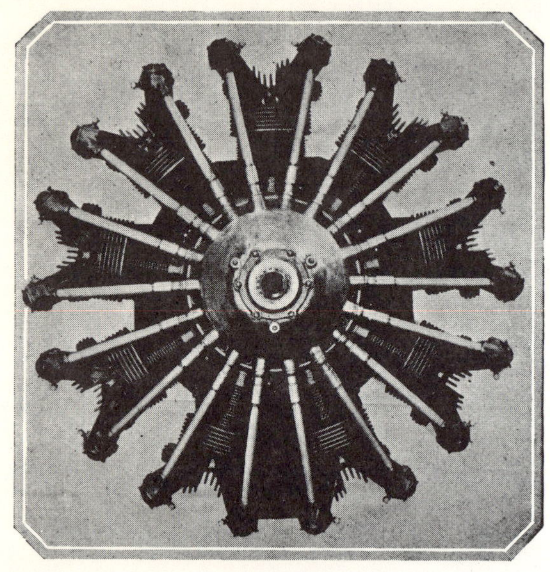

Name of Engine:	**PEGASUS.**
Manufactured by:	Rocky Mountain Steel Products, Inc., 1346-56 Wall St., Los Angeles, Calif.
Type:	9 cylinder, fixed radial, air cooled.
Commercial Rating:	225 H.P. at 1800 R.P.M.
Displacement:	832 cu. in.
Compression Ratio:	5.25 to 1.
Dimensions:	Length, overall .... 40″ Diameter, overall .... 45″ Bore .... 4.625″ Stroke 5.5
Weight:	510 lbs.
Fuel Consumption at Rated H.P.:	Not more than .60 lbs per H.P. hr.
Oil Consumption:	Not more than .025 lbs. per H.P. hr.
Lubrication:	Duplex oil pump, filter.
Ignition:	2 Scintilla Magnetos.
Carburetion:	Stromberg NA T4B adjustable.
Spark Plugs:	2 per cyl.
Price:	On application.

Accessories: (Cost Extra)

Starter.

Generator.

Special Features:

One-piece master rod with tubular articulated rods.

Two-piece crankshaft.

Triple integral induction system.

Fully enclosed and lubricated valve gearing.

Ball bearing rocker arms.

Full tulip valves.

All accessories grouped at rear and pass through 19⅜ mounting ring.

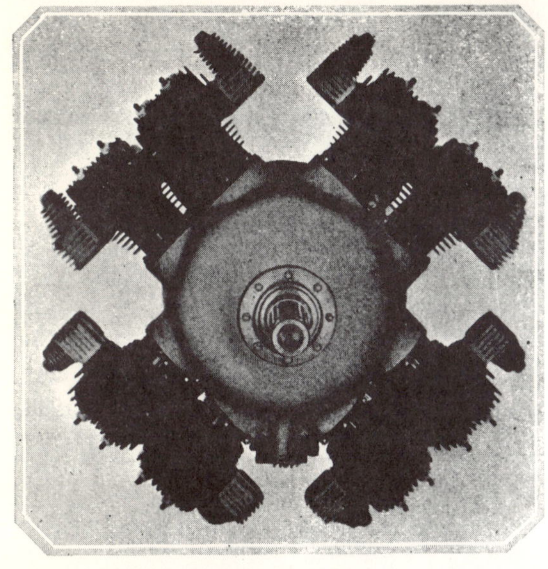

Name of Engine:	**BAKEWELL WING-FOOT.**
Manufactured by:	Shaw-Palmer-Bakewell Co., 409 E. Third St., Los Angeles, Cal.
Type:	Geared V, Radial.
Commercial Rating:	160 H.P. at 2800 R.P.M. motor speed—1800 propeller speed, or can be varied to suit designers' requirements.
Displacement:	456 cu. in.
Compression Ratio:	6 to 1.
Dimensions:	Length, overall 32" Diameter, overall 34" Bore 4" Stroke 4½"
Weight:	Approx. 400 lbs., less starter and propeller hub.
Fuel Consumption at Rated H.P.:	Not more than .55 lb. per H.P. hr.
Oil Consumption:	Not more than .020 lb. per H.P. hr.
Lubrication:	Force feed with splash distribution.
Ignition:	Scintilla.
Carburetion:	Stromberg dual.
Spark Plugs:	2 per cyl.
Price:	$3,500 less hub and starter.

Accessories: (Cost Extra)

Starter—$250. Exhaust stacks—$80. Exhaust manifold—$250.

Special Features:

This Aircraft Motor is of a new design, being constructed so as to be inherently balanced and geared in such a way as to eliminate the usual troubles of gearing; this construction permitting high motor speeds and low propeller speeds. Gear ratios can be furnished to suit customer's requirements.

The motor is equipped with ball-bearings throughout with the exception of connecting rod crank pin bearing which is roller and the wrist pin connecting rod bearing is bronze.

It has been found that this motor does not have a vibrating point up to 4700 R.P.M. and operates very efficiently and quietly.

Crank shafts are of 3 piece construction. Cylinder barrels are Nitralloy steel with the bore hardened to resist wear. Rocker arms being enclosed in crank case are continually lubricated eliminating rocker arm troubles and it has been found that the L-head motor of this small bore and stroke keeps cool very readily at high speeds and has not developed over-heating trouble of any kind.

Valves are operated with single cam for each cylinder, each cam shaft being mounted on two ball bearings. Each crankshaft is mounted on 3 ball bearings and the propeller shaft, being straight thru the motor, is mounted also on three ball bearings.

All oil lines are cast integral with the gear case sections and the oiling system is operated with two hardened gear pumps; one pump removing the oil from the sump and replacing it in the supply tank, the second taking its supply from the tank and forcing it into the propeller shaft where the oil is distributed from this point by force feed splash system. This construction simplifies the oiling mechanism and reduces the possible chance of troubles to a minimum.

This motor is covered by "Patents applied for," on many features.

Name of Engine:	**SIEMENS 5 (SH 13).**
Manufactured by:	Siemens & Halske A. G. (Dr. K. G. Frank, Gen'l Agent), 75 West St., New York, N. Y.
Type:	5-cylinder, radial. air cooled (Approved Dept. of Commerce).
Military Rating:	83 H.P. at 1710 R.P.M.
Commercial Rating:	83 H.P. at 1710 R.P.M.
Displacement:	317 cu. in.
Compression Ratio:	5.3 to 1.
Dimensions:	Length, overall .... 33.8" Diameter, overall .... 39.25" Bore .... 4 133" Stroke .... 4.724"
Weight:	247 lbs.
Fuel Consumption at Rated H.P.:	Not more than .52 lbs. per H.P. hr.
Oil Consumption:	Not more than .027 lbs. per H.P. hr.
Lubrication:	Dry sump. forced feed, low pressure due to complete ball bearing engine.
Ignition:	2 Siemens or Scintilla Magnetos.
Carburetion:	1 Sum Carburetor.
Spark Plugs:	2 per cyl. Siemens.
Price:	$2,100.

Special Features:

Low pressure due to the engine being completely ball bearing.

Cylinders are all of steel barrel with aluminum alloy ribs being cast on by special process, with aluminum alloy head held down by six studs and removable.

Intake ports at rear and exhaust ports on side.

Crankshaft assembly composed of two major castings and front and rear cover plates all of aluminum alloy.

Two-piece single throw crankshaft.

All connecting rods tubular.

Master rod on two ball bearings.

Aluminum alloy pistons with two compression rings and two oils rings in each.

Tulip shaped exhaust and intake valves.

Name of Engine:	**SIEMENS 7 (SH 14).**
Manufactured by:	Siemens & Halske A. G. (Dr. K. G. Frank, Gen'l Agent), 75 West St., New York, N. Y.
Type:	7-cylinder, radial, air cooled (Dept. Commerce Approved Type Certificate).
Military Rating:	113 H.P. at 1720 R.P.M.
Commercial Rating:	113 H.P. at 1720 R.P.M.
Displacement:	443 cu. in.
Compression Ratio:	5.3 to 1.
Dimensions:	Length, overall .... 32" Diameter, overall .... 39¼" Bore .... 4.133" Stroke .... 4.724"
Weight:	308 lbs.
Fuel Consumption at Rated H.P.:	Not more than .52 lbs. per H.P. hr.
Oil Consumption:	Not more than .027 lbs. per H.P. hr.
Lubrication:	Dry sump, forced feed, low pressure.
Ignition:	2 Siemens or Scintilla Magnetos.
Carburetion:	2 Sum or Special Low Sum Carburetor.
Spark Plugs:	2 per cyl. Siemens.
Price:	$2,100.

Special Features:

Low pressure due to the engine being completely ball bearing.

Cylinders are all of steel barrel with aluminum alloy ribs cast on by special process, with aluminum alloy head held down by six studs and removable.

Intake ports at rear and exhaust ports on side.

Crankshaft assembly composed of two major castings and front and rear cover plates all of aluminum alloy.

Two-piece single throw crankshaft.

All connecting rods tubular.

Master rod on two ball bearings.

Aluminum alloy pistons with two compression rings and two oils rings in each.

Tulip shaped exhaust and intake valves.

Name of Engine:	**SIEMENS 9 (SH 12).**
Manufactured by:	Siemens & Halske A. G. (Dr. K. G. Frank, Gen'l Rep.), 75 West St., New York, N. Y.
Type:	9-cylinder, radial, air cooled (Dept. Commerce Approved Certificate).
Military Rating:	128 H.P. at 1736 R.P.M.
Commercial Rating:	128 H.P. at 1736 R.P.M.
Displacement:	517 cu. in.
Compression Ratio:	5.3 to 1.
Dimensions:	Length, overall ....32" Diameter, overall ....40½" Bore .... 3.937" Stroke .... 4.724"
Weight:	Dry, 382 lbs.
Fuel Consumption at Rated H.P.:	Not more than .53 lbs. per H.P. hr.
Oil Consumption:	Not more than .027 lbs. per H.P. hr.
Lubrication:	Dry sump. forced feed, low pressure.
Ignition:	2 Siemens Magnetos.
Carburetion:	2 Sum Carbureters.
Spark Plugs:	2 per cylinder.
Price:	On application.

Accessories: (Cost Extra)

Exhaust Manifold.

Bosch Electric Starter.

Special Features:

Low pressure due to the engine being completely ball bearing.

Cylinders of steel barrel with aluminum alloy head and screwed and shrunk on, also locked by special nut and counter ring. Intake and exhaust ports on the side.

Crankshaft assembly composed of two major castings and front and rear cover plates all of aluminum alloy.

Two-piece single throw crankshaft.

All connecting rods tubular.

Master rod on two ball bearings.

Aluminum alloy pistons with two compression rings and two oil rings on each piston.

Tulip shaped exhaust and intake valves.

Name of Engine:	**SR-3,** Model L.
Manufactured by:	Szekely Aircraft & Engine Co., Holland, Mich.
Type:	3-cylinder, static radial, L head, air-cooled.
Rating:	40 H.P. at 1800 R.P.M.
Displacement:	190 cu. in.
Compression Ratio:	4.8 to 1.
Dimensions:	Diameter, overall .... 29½" Bore .... 4⅛" Stroke .... 4¾"
Weight:	117 lbs. dry. 142 lbs. complete.
Fuel Consumption at Rated H.P.:	Not over .60 lbs. per H.P. hr.
Oil Consumption:	Not over .016 lbs. per H.P. hr.
Lubrication:	Gear type oil and scavenger pumps.
Ignition:	Dual Scintilla magnetos.
Carburetion:	Stromberg, 1½ balanced.
Spark Plugs:	2 per cyl., B. G.
Price:	On application.

Equipment:

Carbureter.
Magnetos.
Propeller Hub.
Oil Pump.
Spark Plugs.
Wires.

Accessories: (Cost Extra)

Starter.

Special Features:

Specially designed for compactness and light weight.
Counterbalanced crankshaft, 2-piece, of Nickel Chrome steel S.A.E. 3140, 1 37/64" diameter.
Cylinders of close-grained Nickel Cast Iron with head cast integral.

Name of Engine:	SR-5, Model L.
Manufactured by:	Szekely Aircraft and Engine Co., Holland, Mich.
Type:	5-cylinder, static radial, L head.
Rating:	70 H.P. at 1800 R.P.M.
Displacement:	315 cu. in.
Compression Ratio:	4.8 to 1.
Dimensions:	Diameter, overall ....31" Bore .... 4⅛" Stroke .... 4¾"
Weight:	186 lbs. complete.
Fuel Consumption at Rated H.P.:	Not more than .60 lbs. per H. P. hr.
Oil Consumption:	Not over .016 lbs. per H.P. hr.
Lubrication:	Gear type oil and scavenger pumps.
Ignition:	Dual Scintilla Magnetos.
Carburetion:	Stromberg, 1½ balanced.
Spark Plugs:	2 per cyl., B. G.
Price:	On application.

Equipment:

Carbureter.
Magnetos.
Propeller Hub.
Oil Pump.
Spark Plugs.
Wires.

Accessories: (Cost Extra)

Starter.

Special Features:

Counterbalanced crankshaft, 2-piece, of Nickel Chrome Steel S.A.E. 3140. 1-25/32" diameter.

Cylinders of close-grained Nickel Cast Iron with head cast integral.

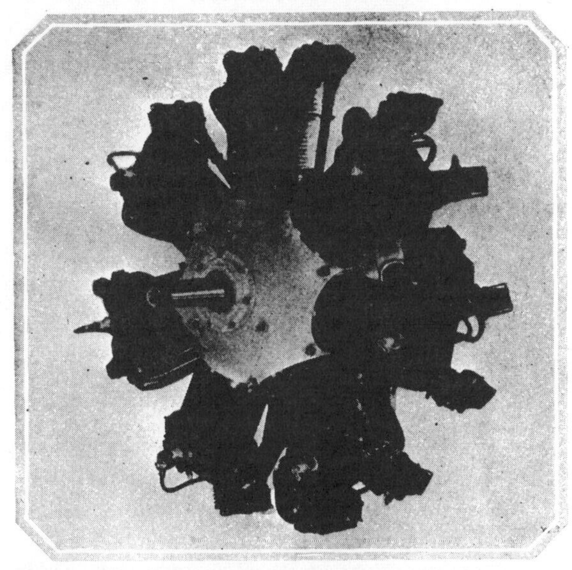

Name of Engine:	**SCARAB.**
Manufactured by:	Warner Aircraft Corp., 20263 Hoover Ave., Detroit, Mich.
Type:	7-cylinder, fixed radial, air cooled, 4-cycle (Approved Type Certificate No. 2).
Military Rating:	110 H.P. at 1850 R.P.M.
Commercial Rating:	110 H.P. at 1850 R.P.M.
Displacement:	422 cu. in.
Compression Ratio:	5.2 to 1.
Dimensions:	Length, overall, without starter, 29" Diameter, overall .... 35½" Bore .... 4¼" Stroke .... 5¼"
Weight:	270 lbs. dry without hub or starter.
Fuel Consumption at Rated H.P.:	Not more than .55 lbs. per H.P. hr.
Oil Consumption:	Not more than .025 lbs. per H.P. hr.
Lubrication:	Force feed.
Ignition:	Dual Scintilla Magnetos.
Carburetion:	1 Stromberg.
Spark Plugs:	2 per cyl. A. C.
Price:	$2,700.

Accessories: (Cost Extra)
Air or Hand-Electric Starter.

Name of Engine:	**WHIRLWIND-165,** Series J-6, Model R-540.
Manufactured by:	Wright Aeronautical Corporation, Paterson, N. J.
Type:	5-Cylinder, fixed radial, aircooled.
Military Rating:	165 H.P. at 2,000 R.P.M.
Commercial Rating:	165 H.P. at 2,000 R.P.M.
Displacement:	540 cu. in.
Compression Ratio:	5.1 to 1.
Dimensions:	Length, overall (no starter) .... 40⅝″ Diameter, overall .... 45″ Bore .... 5″ Stroke .... 5½″
Weight:	Dry (without special equipment) approx. 370 lbs.
Fuel Consumption at Rated H.P.:	Not more than .55 lbs. per H.P. hr.
Oil Consumption:	Not more than .035 lbs. per H.P. hr.
Lubrication:	Pressure pumps.
Ignition:	Scintilla Dual.
Carburetion:	Stromberg, Single Barrel.
Spark Plugs:	2 per cyl.
Price:	$3,000.

Equipment:

Air Cleaner and Heater.
Nose Cowling.
Exhaust Manifold, Complete.
Priming Pump.
Ignition Switch.
Tool Kit.
External Oil Filter.
Instruction Book.

Accessories: (Cost Extra)

Eclipse Hand Inertia Starter.
Eclipse Generator.
Eclipse Generator Control Box.
Standard Steel Propeller Hub.
Fuel Pump.
Eclipse Combination Hand & Electric Inertia Starter.
Eclipse Hand Starter with Booster Magneto
Eclipse Electric Inertia Starter.
Propeller Hub for Wooden Prop.
Metal Prop. Hubs—2 or 3 blade.

Special Features:

Cylinders are composed of a steel barrel over which an aluminum alloy head is screwed and shrunk. Intake ports are at the rear with exhaust ports on the forward side of cylinder.

The crankcase assembly is composed of four major castings of aluminum alloy. The cam follower housing carrying the tappet guides is cast integral with the main section of the crank case.

Two-piece single-throw crankshaft with one-piece master rod and "H" section articulated rods.

Aluminum alloy pistons, cross ribbed on under side of head, and fitted with full floating hollow pins held in place by expanding spring wire locks.

Tulip shaped valves; solid stem inlet valves and hollow stem exhaust valves.

Rotary induction system of Wright design, with "pre-heating" device on carburetor.

Provision is made for obtaining, at all times, a clean supply of air to the carburetor.

Lubrication system is designed to eliminate all external oil pipes to engine.

Name of Engine:	**WHIRLWIND-225, Series J-6, Model R-760.**
Manufactured by:	**Wright Aeronautical Corporation, Paterson, N. J.**
Type:	**7-cylinder, fixed-radial, aircooled.**
Military Rating:	**225 H.P. at 2,000 R.P.M.**
Commercial Rating:	**225 H.P. at 2,000 R.P.M.**
Displacement:	**756 cu. in.**
Compression Ratio:	**5.1 to 1.**
Dimensions:	**Length, overall (no starter) .40-23/32"** **Diameter, overall ....45"** **Bore .... 5"** **Stroke .... 5½"**
Weight:	**Dry (without special equipment) approx. 445 lbs.**
Fuel Consumption at Rated H.P.:	**Not more than .55 lbs. per H.P. hr.**
Oil Consumption:	**Not more than .035 lbs. per H.P. hr.**
Lubrication:	**Pressure Pumps.**
Ignition:	**Scintilla Dual**
Carburetion:	**Stromberg, Single Barrel.**
Spark Plugs:	**2 per cyl.**
Price:	**$3,900.**

Equipment:

Air Cleaner and Heater.
Nose Cowling.
Exhaust Manifold, Complete.
Priming Pump.
Ignition Switch.
Tool Kit.
External Oil Filter.
Instruction Book.

Accessories: (Cost Extra)

Eclipse Hand Inertia Starter.
Eclipse Generator.
Eclipse Generator Control Box.
Standard Steel Propeller Hub.
Fuel Pump.
Eclipse Combination Hand & Electric Inertia Starter.
Eclipse Hand Starter with Booster
Magneto
Eclipse Electric Inertia Starter.
Propeller Hub for Wooden Propellers.
Metal Propeller Hubs—2 or 3 Blade.

Special Features:

Cylinders are composed of a steel barrel over which an aluminum alloy head is screwed and shrunk. Intake ports are at the rear with exhaust ports on the forward side of cylinders.

The crankcase assembly is composed of four major castings of aluminum alloy. The cam follower housing carrying the tappet guides is cast integral with the main section of the crank case.

Two-piece single-throw crankshaft with one-piece master rod and "H" section articulated rods.

Aluminum alloy pistons, cross ribbed on under side of head, and fitted with full floating hollow pins held in place by expanding spring wire locks.

Tulip shaped valves; solid stem inlet valves and hollow stem exhaust valves.

Rotary induction system of Wright design, with carburetor-air preheating device.

Provision is made for obtaining, at all times, a clean supply of air to the carburetor.

Lubrication system is designed to eliminate all external oil pipes on engine.

Name of Engine:	**WHIRLWIND-300,** Series J-6, **Model R.975.**
Manufactured by:	Wright Aeronautical Corp., Paterson, N. J.
Type:	9-Cylinder, fixed radial, aircooled.
Military Rating:	300 H.P. at 2,000 R.P.M.
Commercial Rating:	300 H.P. at 2,000 R.P.M. (sea level)
Displacement:	972 cu. in.
Compression Ratio:	5.1 to 1.
Dimensions:	Length, overall .... 41-7/16" Diameter overall .... 45" Bore .... 5" Stroke .... 5-1/2"
Weight:	Dry, without special equipment, approx. 520 lbs.
Fuel Consumption at Rated H.P.:	Not more than .55 lbs. per H.P. hr.
Oil Consumption:	Not more than .035 lbs. per H.P. hr.
Lubrication:	Pressure pump.
Ignition:	Scintilla, Dual.
Carburetion:	Stromberg, Single bbl.
Spark Plugs:	2 per cyl.
Price:	$4,800.

Equipment:

Air cleaner & heater
Nose Cowling
Exhaust Manifold, complete
Priming Pump
Ignition Switch
Tool Kit
External Oil Filter
Instruction Book

Accessories: (Cost Extra)

Eclipse Hand Inertia Starter
Eclipse Generator
Eclipse Generator Control Box
Standard Steel Propeller Hub
Fuel Pump
Metal Prop. Hubs—2 or 3 blades
Eclipse Combination Hand & Electric Inertia Starter
Eclipse Hand Starter with Booster Magneto
Eclipse Electric Inertia Starter
Propeller Hub for Wooden Prop.

Special Features:

Cylinders are composed of a steel barrel over which an aluminum alloy head is screwed and shrunk. Intake ports are at the rear with exhaust ports on the forward side of cylinders.

The crank case assembly is composed of four major castings of aluminum alloy. The cam follower housing carrying the tappet guides is cast integral with the main section of the crank case.

Two piece, single throw crank shaft with one piece master rod and "H" section articulated rods.

Aluminum alloy pistons, cross ribbed on under side of head, and fitted with full floating hollow pins held in place by expanding spring wire locks.

Tulip shaped valves; solid stem inlet valves and hollow stem exhaust valves.

Rotary induction system of Wright design, with carburetor-air preheating device.

Provision is made for obtaining, at all times, a clean supply of air to the carburetor.

Lubrication system is designed to eliminate all external oil pipes on engine.

Name of Engine:	**CYCLONE-NINE,** Model R-1750.
Manufactured by:	Wright Aeronautical Corporation, Paterson, N. J.
Type:	9-cylinder, fixed-radial, aircooled.
Military Rating:	525 H.P. at 1,900 R.P.M.
Commercial Rating:	525 H.P. at 1,900 R.P.M.
Displacement:	1,750 cu. in.
Compression Ratio:	5:1.
Dimensions:	Length, overall (no starter) .39¼" Diameter, overall ....53-15/16" Bore .... 6" Stroke .... 6⅞"
Weight:	Dry (without special equipment) approx. 770 lbs.
Fuel Consumption at Rated H.P.:	Not more than .60 lbs. per H.P. hr.
Oil Consumption:	Not more than .035 lbs. per H.P. hr.
Lubrication:	Pressure Pumps.
Ignition:	Scintilla Dual.
Carburetion:	Stromberg, double barrel.
Spark Plugs:	2 per cyl.
Price:	$8,600 direct drive as specified above. $10,000 with 2 to 1 reduction gearing but without nose cowling & exhaust manifold.

Equipment:

Air Cleaner and Heater.
Nose Cowling.
Exhaust Manifold Complete.
Priming Pumps.
Ignition Switch.
Tool Kit.
External Oil Filter
Instruction Book

Accessories: (Cost Extra)

Eclipse Combination Hand & Electric Inertia Starter.
Eclipse Hand Starter with Booster Magneto.
Eclipse Electric Inertia Starter.
Propeller Hub for Wooden Propellers.
Metal Propeller Hubs—2 or 3 Blade.
Eclipse Hand Inertia Starter.
Eclipse Hand and Electric Inertia Starter.
Switches for Electric Inertia Starter.
Eclipse Generator.
Eclipse Generator Control Box.
Fuel Pump.
Standard Steel Propeller Hub (3-bladed).

Special Features:

Cylinders are composed of a steel barrel over which an aluminum alloy head is screwed and shrunk. Intake ports are at the rear with exhaust ports on the forward side of cylinders.

Crankcase assembly is composed of five major castings of aluminum alloy. The cam follower ring carrying the tappet guides is cast integral with the main section of the crankcase.

Two-piece single-throw crankshaft with one-piece master rod and "I" section articulated rods.

Aluminum alloy pistons, cross ribbed on the under side of the head, and fitted with full floating hollow pins.

Tulip shaped valves; solid stem inlet valve and hollow stem exhaust valve; the exhaust valve containing a special salt compound which improves valve cooling and prevents the warping usually encountered in valves of this large size.

Name of Engine:	**WRIGHT GIPSY.**
Manufactured by:	Wright Aeronautical Corp., Paterson, N. J.
Type:	4-Cylinder, aircooled, vertical-in-line.
Military Rating:	85 H.P. at 1,900 R.P.M.
Commercial Rating:	100 H.P. at 2100 R.P.M.
Displacement:	318 cu. in.
Compression Ratio:	5 to 1.
Dimensions:	Length, overall, including propeller hub ....45¼" Width between mounting pads ....11-11/16" Bore .... 4½" Stroke .... 5"
Weight:	Dry, including standard equipment, approx. 285 lbs.
Fuel Consumption at Rated H.P.:	Not more than .55 lbs. per H.P. hr.
Oil Consumption:	Not more than .010 lbs. per H.P. hr.
Lubrication:	Pressure (Gear Pump). No scavenger pump.
Ignition:	Scintilla, Dual, Type PN-4-D. (Impulse starting device on one.)
Carburetion:	Stromberg Type NA-R-4A.
Spark Plugs:	2 per cyl.
Price:	$1,600.

Equipment:

Cylinder Cowling
Exhaust Manifold
Ignition Switch
Tool Kit
Instruction Book.

Accessories (Cost Extra)

Starter, Eclipse Hand.
Propeller Hub, Flange and Bolts.

Special Features:

Nickel Cast Iron cylinders. Detachable aluminum alloy cylinder heads with one intake and one exhaust valve each and extruded bronze valve seats shrunk into heads.

Crankshaft 1 piece forging, 5 bearings.

Tulip type valves of special valve steel.

Aluminum alloy pistons, slipper type. Hot spot effect secured by passing that portion of the intake manifold immediately above the carburetor thru an exhaust manifold jacket.

Name of Engine:	W-T-5.
Manufactured by:	Wright-Tuttle Aircraft Motors Corp., West 25th and Walton Sts., Anderson, Ind.
Type:	5-cylinder, fixed radial, air-cooled (Dept. of Commerce Approved Type).
Rating:	100 H.P. at 1400 R.P.M.
Displacement:	482 cu. in.
Compression Ratio:	5.2 to 1.
Dimensions:	Length, overall ....34 9/16" Diameter, overall ....42.5" Bore .... 4¾" Stroke .... 5½"
Weight:	275 lbs. complete.
Fuel Consumption at Rated H.P.:	Not more than .503 lbs. per H.P. hr.
Oil Consumption:	Not more than .022 lbs. per H.P. hr.
Lubrication:	Dry sump., one lubricating and one scavenging pump.
Ignition:	Dual Scintilla Magnetos.
Carburetion:	1 Stromberg.
Spark Plugs:	2 per cyl. Champion.
Price:	$1,875.

Special Features:

Simple construction—less than 200 parts.

Light Weight—2.2 lbs. per H.P.

Dual ignition timing 0 to 35 degrees.

Ingenious valve operating mechanism with duraluminum push rods, and rocker arms with ball bearings.

INDEX

AVIATION
PUBLICATIONS

P. O. BOX 123
MILWAUKEE
WISCONSIN
53201
U.S.A.